THE WORLD OF OCEAN ANIMALS

# SEA ANEMONES

by Adeline J. Zimmerman

pogo

# Ideas for Parents and Teachers

Pogo Books let children practice reading informational text while introducing them to nonfiction features such as headings, labels, sidebars, maps, and diagrams, as well as a table of contents, glossary, and index.

Carefully leveled text with a strong photo match offers early fluent readers the support they need to succeed.

## Before Reading

- "Walk" through the book and point out the various nonfiction features. Ask the student what purpose each feature serves.
- Look at the glossary together. Read and discuss the words.

## Read the Book

- Have the child read the book independently.
- Invite him or her to list questions that arise from reading.

## After Reading

- Discuss the child's questions. Talk about how he or she might find answers to those questions.
- Prompt the child to think more. Ask: Did you know about sea anemones before reading this book? What more would you like to learn?

Pogo Books are published by Jump!
5357 Penn Avenue South
Minneapolis, MN 55419
www.jumplibrary.com



Library of Congress Cataloging-in-Publication Data

Names: Zimmerman, Adeline J., author.
Title: Sea anemones / Adeline J. Zimmerman.
Description: Minneapolis: Jump!, Inc., [2022]
Series: The world of ocean animals
Includes index. | Audience: Ages 7-10
Identifiers: LCCN 2021021722 (print)
LCCN 2021021723 (ebook)
ISBN 9781636902913 (hardcover)
ISBN 9781636902920 (paperback)
ISBN 9781636902937 (ebook)
Subjects: LCSH: Sea anemones–Juvenile literature.
Classification: LCC QL377.C7 Z56 2022 (print)
LCC QL377.C7 (ebook) | DDC 593.6–dc23
LC record available at https://lccn.loc.gov/2021021722
LC ebook record available at https://lccn.loc.gov/2021021723

Editor: Jenna Gleisner
Designer: Michelle Sonnek

Photo Credits: William Drumm/Shutterstock, cover; Josephine Jullian/iStock, 1; Cloudpost/Shutterstock, 3; Joe Belanger/Shutterstock, 4; Michael Zeigler/iStock, 5; RLS Photo/Shutterstock, 6-7; Pally/Alamy, 8-9; Auscape International Pty Ltd/Alamy, 10-11; Boris Pamikov/Shutterstock, 12; aquapix/Shutterstock, 13; Jesus Cobaleda/Shutterstock, 14-15tl; blickwinkel/Alamy, 14-15tr; Randy Bjorklund/Shutterstock, 14-15bl; Vojce/Shutterstock, 14-15br; Rich Carey/Shutterstock, 16; Gialdini Luca/Shutterstock, 17; bayazed/Shutterstock, 18-19; valda butterworth/Shutterstock, 20-21; Aaron Bull/iStock, 23.

Printed in the United States of America at Corporate Graphics in North Mankato, Minnesota.

# TABLE OF CONTENTS

# CHAPTER 1

# PRETTY PREDATORS

This colorful sea creature may look like a flower. But it is an animal! What is it? It's a sea anemone! Sea anemones are **invertebrates**. They don't have backbones. In fact, they don't have any bones at all!

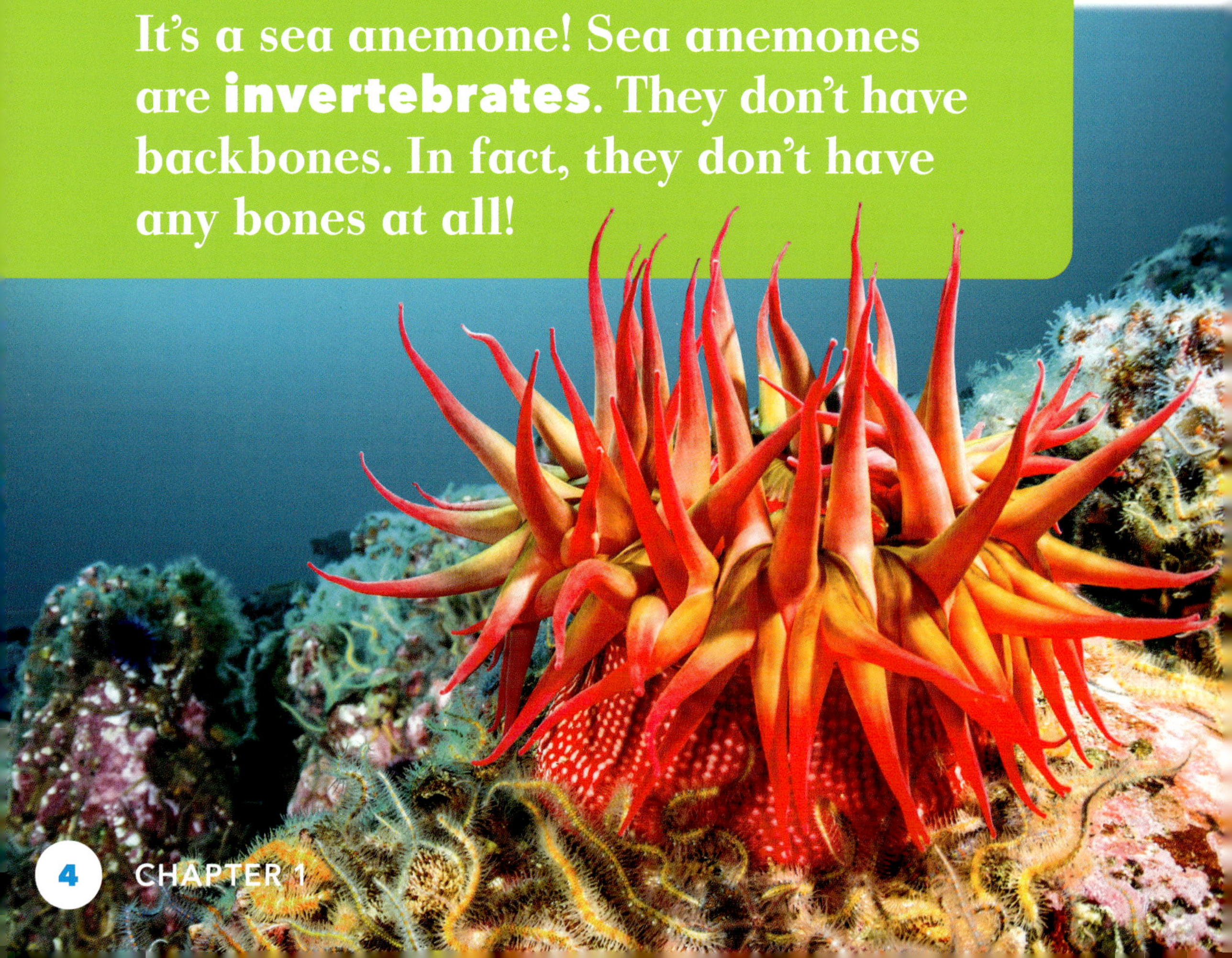

They are **cnidarians**, like corals and jellyfish. They do not have brains. But they are smart **predators**. They use stinging **tentacles** to catch **prey**.

Tentacles circle each anemone's mouth. Their bodies are shaped like tubes. Some are short and thick. Others are long and thin. They attach their bases to rocks and other hard surfaces.

# TAKE A LOOK!

What are a sea anemone's body parts called? Take a look!

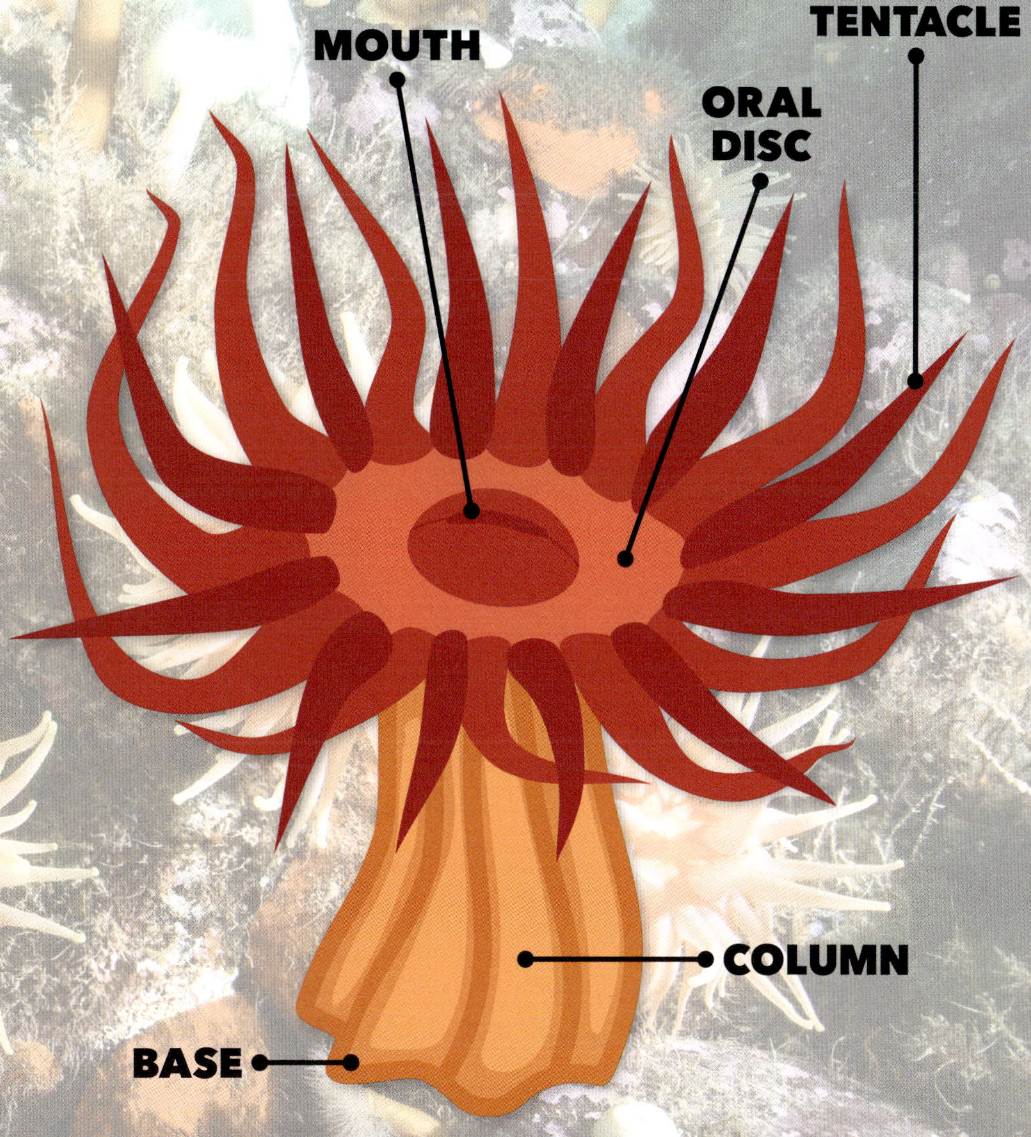

Sea anemones often stay in one spot. They wait for food to come to them. They are **carnivores**. They eat mostly small crabs, fish, and **plankton**.

When prey swims by, it touches the tentacles. These sting the prey so it cannot move. Then they move the prey into the anemone's mouth. Yum!

**DID YOU KNOW?**

Larger sea anemones eat larger prey. The dahlia anemone is one. It eats sea stars and jellyfish!

While sea anemones often stay in one place, they can move and swim. How? They glide on their bases. They can also **detach**. They catch a **current**. Then they attach somewhere else.

## DID YOU KNOW?

Why might sea anemones need to move? The biggest reason is predators. Sea slugs, sea stars, eels, and fish all eat anemones.

# CHAPTER 2

# KINDS OF SEA ANEMONES

**Sea anemones are found throughout the world's oceans. Some live more than 32,000 feet (9,750 meters) deep.**

Many live in warm, **tropical** waters. This is where the largest and most colorful anemones are found.

There are more than 1,000 sea anemone **species**. They come in many colors, sizes, and shapes. Take a look at these anemones. Can you guess how they got their names?

**DID YOU KNOW?**

The giant green anemone is one of the largest. It can be almost three feet (1.0 m) across. That is as wide as a large refrigerator!

bubble-tip anemone

long tentacle anemone

giant green anemone

carpet anemone

# CHAPTER 3

# SEA ANEMONE RELATIONSHIPS

Sea anemones have some interesting **relationships** with other animals. One is with clownfish. These bright fish live among the tentacles of some anemones. They stay safe from predators.

A layer of **mucus** protects the fish. This way, the tentacles don't sting it. The relationship helps the anemone, too. How? It eats food scraps the clownfish leaves behind.

Some crabs carry sea anemones on their backs. The stinging tentacles help protect the crab. They are also good **camouflage**! In return, the anemone gets scraps of food while the crab eats.

Some anemones live in groups called **colonies**. Members protect each other and their space. If an anemone from a different colony gets close, a warrior anemone slaps it. This leaves stinging cells on the unwelcome anemone.

Sea anemones are interesting creatures! Would you like to see one?

**DID YOU KNOW?**

Each sea anemone colony has a scout. This anemone looks for new space to claim.

colony

# ACTIVITIES & TOOLS

## TRY THIS!

### SEA ANEMONE TAG

Sea anemone colonies take over new spaces and fend off intruders. Play this fun game of tag to see how their colonies work!

**What You Need:**

- a group of friends or classmates
- a room with objects to move around

1. Split the group into two "colonies."
2. Have each colony stand together on opposite sides of the room.
3. At the same time, have each group make its way to the opposite side of the room. Spread out a little as you walk, but make sure to stay together. Look for new places for your colony to "stick" to.
4. If a member from the other colony gets close, reach out a tentacle (your hand) and tag that anemone! How many anemones from the other colony did your colony tag on its way across the room?

# GLOSSARY

**camouflage:** A disguise or natural coloring that allows animals to hide by making them look like their surroundings.

**carnivores:** Animals that eat meat.

**cnidarians:** Aquatic animals that are invertebrates and have tentacles.

**colonies:** Groups of sea anemones.

**current:** The movement of water in one direction.

**detach:** To separate from something.

**invertebrates:** Animals that do not have backbones.

**mucus:** A thick, slimy liquid that coats and protects.

**plankton:** Tiny animals and plants that drift or float in oceans and lakes.

**predators:** Animals that hunt other animals for food.

**prey:** Animals that are hunted by other animals for food.

**relationships:** The ways in which two or more things are connected.

**species:** One of the groups into which similar animals and plants are divided.

**tentacles:** The flexible limbs of some ocean animals.

**tropical:** Of or having to do with the hot, rainy area of the tropics.

## INDEX



## TO LEARN MORE

Finding more information is as easy as 1, 2, 3.

1. Go to www.factsurfer.com
2. Enter "seaanemones" into the search box.
3. Choose your book to see a list of websites.

FACT SURFER